TRUTH **SERUM** PRESS

MY LIFE
IN CARS

POEMS BY
ED RUZICKA

PHOTOGRAPHY BY
CHARLES N. deGRAVELLES

First published as a collection November 2020
Poems copyright © Ed Ruzicka
Car photographs copyright © Charles N. deGravelles

BP#00098

ISBN: 978-1-922427-10-6

Truth Serum Press
32 Meredith Street
Sefton Park SA 5083
Australia

Email: truthserumpress@live.com.au
Website: https://truthserumpress.net/
Store: https://truthserumpress.net/catalogue/

Front cover photograph copyright © Charles N. deGravelles,
used with permission
Cover design copyright © Matt Potter

Also available as an eBook:
ISBN: 978-1-922427-11-3

Truth Serum Press is a member of the
Bequem Publishing collective
http://www.bequempublishing.com/

This book is dedicated
to my wife

Renee Stickels Ruzicka

whose tolerance, dedication and affection
stun me daily and help me strive to
move on up a little higher.

I'm going home one day, tell my story
I been climbing over hills and mountains ...
Gonna drink from that old healing fountain.

I Will Move on Up a Little Higher
by W. Herbert Brewster
as sung by Mahalia Jackson

Cars are America.

Marty Marak

The Gods with unpronounceable names ... want
the beating hearts from cars
to be head high on the altars.

Speech Scroll (98) by David Chorlton

Special thanks go to

Jane Napolitano

and my eldest daughter

Helen Miriam

for spending hours correcting
oversights, typos and blunders

and

Kel LeJeune

of Kel's Custom Classics, LLC
for allowing Charles and me into his shop
to photograph his handiwork.

CONTENTS

AN OLDSMOBILE, A PACKARD

INSATIABLE THROAT

GOING THROUGH THE GEARS

TIME AND THE WHEEL

CARS OF MEMORY AND FORGETTING

An Oldsmobile, a Packard

My Engines

When I was a boy beside cornfields in a Buick Eight
I'd put my hand out the open window, palm flat,
extended, to surf furious, wrenching air.

That was one way to acquaint myself
with grace in the presence of power.
I had no power but surges of imagination
within the machinations of boredom.

I had a wild stallion that raced
along the borders of flung fields.
Effortlessly galloped beside tree lines,
leaped fences. Able to keep pace
with the hundreds of horsepower
under our Buick's hood.

I had a scythe that others could not see
that sliced down what we passed.
It all fell behind us. Only I knew how weed,
bush, sapling bowed for this boy's passing.

I had nothing and was skinny as the posts
that weathered in the sun grey to silver –
as my hair is now. Posts held barbed wire menace
that a boy could also pick his way through
headed down to the river or off into woods.

Later, I went thousands of places in hundreds of cars
that took me to jobs, carried groceries,
gave enough privacy for women to slip
beneath me in the back seat.
Eventually, I didn't need turns of fantasy,
did not need a boy's lust for power,
could often hold what I wanted.

I reigned in the mighty steeds of imagination
strong as fire round the sun. Now I can hammer, saw, weld,
or write things that last, things that do not bow to anyone.
I stand wild, solid and raging where I live
between memory and imagination.

What Dad Said While He Drove

Willow leaves spindle, curl
as they yellow, brown, crinkle.
Willow branches fall slender.
Whip ends whisk, swirl –
horse tails in a storm.

Roots seek, draw water,
send a fountain up from trunk,
that strikes a silhouette,
shivers silver in full moon.

Willows line river banks where
untended kids and lovers drift.
Reflect sideways in lake shimmer.
Nurse shadows in their skirts.

My dad had a game, had a
contest while we drove, a diversion.
Who could count the most willows
before we could get to where we went.

The rest of the time Dad was prosaic:
work, make hay while the sun shines,
the Republican party, the religion
of major league baseball.

When he held the wheel he'd say –
and I would wait, wait, wait for it to come –
"If a tree can be said to have gender
then surely the willow is feminine."

WHEN TOWNS WORKED

Dad'd drum his pencil rapid on the counter,
say, "Let me find a boy scout to take you home."
He'd wait up there on the mezzanine platform
till the right one showed. Bellow a name.
Take my shoulder, shove me off with an
Earl or a Ray. Sometimes slip a candy bar
in my palm for the ride. I'd go out, February
wind a knife between jacket buttons.
Load into a pickup, an Electra, a bronze Impala,
Feel gears mesh, vibrate through floorboard.
Most didn't bother talking, let fields roll.

Some would ask, Eddie this? Eddie that?
as I tried to find a place for my feet in a squall
of receipts, pipes, wrenches. Or as I saw
a stag race past trees along the river,
frozen then hard as any highway.
After laying tracks in driveway snow
they'd let me out with a hand gesture
as my black and white pup danced up.

What I took away was the gentle cynicism
of Dad's phrase, "boy scout", when later that evening
he would send an employee out in the Scout International
to a neighboring town where a former Warrenville
resident had moved to an apartment, fallen
on hard times, left behind fields
and a frame-house punished by blizzards.

Dad would tell the delivery boy, "Don't you let them
give you a dime for this." – prescription
in a white bag creased at the top.
He'd think nothing of it again. Wind.

Two Box Loads of Sunday Pastry

I

Sometimes Dad woke me, to take me into
the nailhead of winter, over iced roads
in his white Corvair. Headlights tunneled into snow,
flew over roadside gravel, as I tried to find
dream again by burrowing into the seat-back.

There was nothing for a boy all the way along
Butterfield Road 'til the car lurched right onto Madison
where brick facades and stop lights began at regular
intervals; street lamp light galvanized concrete.

Left onto Harlem Ave. four blocks. Ease tires to curb,
crusted snow, slush blackened by cinder and exhaust.
I sat as Dad hopped out, secured his top coat
and tucked his chin to muscle toward the bakery.

Strong blasts of wind scuttled newsprint up the street.
Flung it into tiny cyclones within trapezoid-shaped
storefronts. Each store glass-encased and reinforced
with metal cage-works that the jeweler or the cleaner
would unlock weekdays while light was still bare,

electrical and grim. But this was Sunday morning.
The day Dad drove over thirty miles to get the week's
finest confectionery treats from a bakery not far from
where he'd spent his boyhood. Not that far from where
his father's own "Apothecary and Pharmacy" once stood.

II

Dad called his fountain at the back of the store,
"the talk of the town" because everybody came in
 for Coke, seltzers, coffee, and everybody gabbed.
January Sundays were bustling affairs that began
with a raucous row of plaid-jacketed fox hunters who had just
roamed over fields, along creek banks. Their boots
dripped snow-melt. Having fully escaped dull jobs and duller wives
their profane voices almost broke into song as they greedily
knocked back hot coffee, shoved jelly rolls into their maws.

It was a thirty mile drive to Oak Park for pastry,
thirty back for a six a.m. open. As farmers idled
at the curb, Dad jammed his key into the lock. Twisted.
Flicked fluorescent lights to a low hum. I set
cinnamon rolls, french twists, long johns
onto platters in a display case. Got "a sling a mud"
for Mr. Mason. "Gimme some of that river water you
got there," said Haley who ran the John Deere store.

III

My biggest Sunday job was to stuff the news section
of the Sunday papers with the comics-encased advertisements
and the Sunday Parade Magazine, all of which were printed
by Friday. Around two hundred forty of these, bundled,
bound by wire. A pound and a half of paper each.
Twenty to a pack. Good heft for a nine year old.

Took awhile. Had to stop whenever someone
showed up at the front counter for Chesterfields,
Prince Albert, Witch Hazel, or a fancy lettered box
of chocolate-covered Turtles for his mother's birthday.
Making change, customer's growls taught me math fast.

IV

One Sunday in Oak Park, I guess I got antsy. Got out
of the car. Huge icicles hung down the side of gutter-pipe.
I stumbled over snow but came to an abrupt stop
when a monstrous figure, all meat in the face, erupted
from under a makeshift bed of cardboard
in one of those store fronts. Such a blast of ethanol and piss,

I gagged when he caught my arm. Dragged me
in close. Snarled, "You got anything for an old man,
boy? Got some change, huh." He grabbed up what
few coins I dug out. Then let me scamper off. Fall
to ice. Dust my butt and make it into that bakery

overflowing with the wondrous scent of oven-warmed dough.
When we came out Dad almost bumped into the man
that now muttered and staggered down Harlem St.
My kind-hearted dad gave that man
a look fiercer than I had ever seen.

V

I thought about this when I first found out
that Dad took over his father's drugstore sometime
in the thirties when his parents divorced
and his father left the store, left his family
to pursue a full time position in alcohol.
Ended up with a rancid liver, in rigor mortis
on a flop house floor on South Halstead two decades later,
just about the time I showed up, grey, wet
and crying my fresh-formed lungs out.

Dad would have seen, in that drunk bum's
faltering gait, every flaw his father showed.
Dad took my arm gently but firmly. Guided me
toward the car. Guided me away from everything
his father was, toward his drug store, toward
what he had made of himself, toward the counter
where he would unfold his newspaper, set his
coffee off to the left. There might not
have been enough sugar in those pastries
to take the bitter back out of his days.

Black-eyed Susans, a Stallion, Crows

If God wanted to crush, then revive a man
would he put him in a Corvair as a tree-twisting wind
whistled up and down the road to Naperville?
Would he make the man fly through predawn murk
on Goodyear tires with a V-6 behind him
where the trunk should be and at an hour when only
farmers were up to lumber toward barns?

The fox was nearby startling creek-side creatures
with its nimble weave, field mice nosed under wheat,
snakes slunk and the crows – well they never close
both eyes anyway because death and hunger never rest.

Would God also have let posts rot in the corner
of a field where wind knocked down an ash
so that a black stallion could mince over barbed wire
fallen face down into black-eyed Susans and milkweed,
so a stallion could graze roadside, then trot into headlights?

The man at the wheel whistled with that wind.
He had left his house of five children, the wife who
had grown tired of his agitations. Whether the man
ever had an affair only he could really say
but his wife saw signs enough to throw this up
whenever they argued. The man's eyes were still bleary
when the Corvair took out the legs of the stallion.

The full tonnage of the horse lifted to skid
over the hood and half way through auto glass.
The horse's ribs cracked, metal cracked. Roof
and framing broke downward to the man's forehead.

All stops cold, goes black as black wings.
Except that the farmer now runs to the road.
By the time officers arrive the farmer has the man
        propped against the car. Intermittently the man lifts
        a rag from his forehead to mop a blinding red flow.
For days family and town's-folk talk as if God
        had sent a dozen angels down at just
        the right instant to arrest that farmer's
        horse where it lay on the front of the roof.

My father walked away. My father
came back home with broken ribs,
superficial cuts, a very red shirt.

I believe that whatever God is stayed,
as always, in the depths of earth's core,
the immovable mover, molten and remote
to work his unfathomable math, the strict
laws of his physics and a dizzying wonder
using only the crows for his eyes.

THE MEAN
GREEN MACHINE

My brother's Packard rattled glass panes
warming up. Filled its park space. Was a
glinting green, had hips, twin duck tails
beside the trunk, swoops of tin, sheets of steel
over engine muscle. Seemed to breathe just sitting.

Pure thrill to throttle, warm up. Could fit a
full infield when he picked me up at the ball field.
Had a motel-load of cushion in both back and front
seats where local lasses loosed, shucked, lifted skirts
for our hero-lad lately back from the U. S. Air Force.

Svelte model designed in Hollywood for men in tuxes,
women trailing silks and chiffon but had made its way
down the automotive food chain to a used car lot
in Wheaton, Illinois where my brother borrowed
a couple hundred from Dad to get those keys.

Rode it deep into the hard drinking dark of the Midwest.
The loose cannons of two of our springs, the crowded stars
of two summers, one autumn's turn and a single winter before it
came to broadsided against an eight foot mass of plowed snow.

Still rumbled on, not quite as glorious, under the able hands
of handsome Jerry. Knew the night, that car.
Seemed designed for darkness crowded with dreams
and admired by dreamers, which was all I was, while my brother
ripped past corn fields flying easy as a God of old.

My First Car

came from my brother after he took a brush
and, I swear to God, swiped his Karmann Ghia
with a brilliant, lipstick red paint that caked out
to monkey-shit brown when it dried.
The next morning, he took one look, gave me the keys.
It was late spring. The radio was out so I propped a transistor
up on the console, let wind rip though lyrics.

The seats were split, foam bulged. I didn't care.
The Ghia sizzled and spat all spring and summer
but then something started to happen.
The brake lining took on air – slowed it down
the way bumper cars give out when a carny turns off the juice.

Mechanics I could afford would fix it well enough
to last a month or two. Then it would go back
to where the best thing was to take back roads:
down shift, pump the emergency brake –
totter to a halt at stop signs.

That is what I had to do the winter and spring
of '69 if I missed the bus. I'd ride beside scrub,
rows of corn stubble with the window open
chugging along slow as a tractor. Had to poke
my head out when the windshield was glazed with ice,
hope no country fool pulled out. Back then America
was just starting to use drugs and parents didn't
have time to worry about what their kids were up to.

That summer, just before I joined thousands of other teens
to hitchhike out of blank-slate America to Southern California
in order to see exactly what Jim Morrison was singing about,
I sold the Karmann Ghia for ten dollars to a thugish kid I used
to fist-fight all through Junior High. Ten dollars and it still
ran any time you wanted to risk it. That was my Senior year
when I tooled all over Dupage County looking pretty much
like what I was – a half-cured turd on wheels.

EOLA

There was a gravel road with a bridge over a rail yard
stuck two miles out from Eola, Illinois, which was itself
just a patchwork of fields, an elementary school, P.O.,
a steepled church, and a dozen wood-frame homes.

I used to tell Dad that I was heading out for Sunday mass
and drive there because freedom was my religion.
The bridge was made of bolted, welded steel
crossed with heavy wooden planks that rattled
loud and loose as my tires slowly rolled.

The whole bridge clattered. Wind keened until
you felt just how much air there was below,
falling to rails, to parked box cars – hulking,
rusted, their wheels buried in snow drifts.

At night I'd take girls out. Pump us up with fear
hoping to make it to the back seat together
where only frozen corn-stubble could witness
what we gained or lost. Those days any moon
laying on snow shivered, lean and hungry.

Teen Ode

Those roads had given themselves up
hours before. Lay ghosted, hollow,

except for the rip of our wake.
Mailboxes, milkweed, barbed wire.

It didn't matter whose car we had,
who was driving. Desire was the fuel.

Highways were our veins. Cigarettes
and beer were commas between the

hours and the aching and the movement.
Girls just as aching, broken dolls. Husks

from which women were emerging. We
needed to let ourselves go within them

or they needed us, our needing them. We
did need them and little else but the aching,

the emptiness, and the point of balance
a star's spear could reach as we gazed down

from a rock above quarry waters, ready to
dive head first, again and again, into darkness.

A Very Big Car,
Then a Small One

My sister's first car was an Olds or a Buick
that tilted on shot struts and shocks rounding
any street corner. Listed like a freighter, it did.
My father got it gratis from a customer after
the woman's husband blew his brains out at the wheel.

There was a head-high bullet hole in the window,
though the glass was barely disturbed. Just left
an irregular circle of crumbled fragments clustered
around the car's glass exit wound . Tiny, ice-clear
cubes that somehow didn't blow off even at 60.

Dad handed us the keys and an address. Told us
not to disturb anyone in the house. It was November.
There was a rusted swing and patches of snow
in the yard. The car turned over like a charm before
I noticed my shoes were in a sort of gummy, thin,
jell-O. The man's purple plasma was still pooled
in the well in front of the rider's side seat.

We gladly paid a car wash out on Route 59
twenty bucks to get that out. That beast held up
at least six months before she bought a baby-blue
V. W. Bug and bumbled out on I -80 for Baltimore
and the rest of her life. Didn't come home
for three years more. Came at Christmas. Some punk
in the big city had spray-painted "FUCK"
on the car's side. Clare didn't have enough bucks

to get her Bug repainted or thought it funny.
So she left it. My mother was not amused,
was aghast. The rest of us couldn't care less,
just wanted our bones to vibrate up
from the floor board as fence posts flew
backwards in the rearview mirror.

We just wanted to go anywhere. Accelerate.
"FUCK", hell yea, fuck our way into any
tomorrow where there was more
than a bowling alley, four churches, three bars,
a grocery and enough rows of corn
to feed a murder of crows for infinity.

Go, Just Go

The summer my sister Clare got home from four years
in a liberal arts curriculum at the University of Budweiser,
I had a driver's license and a job at the county airport.
Seated on top of a tractor in 85-degree heat, I mowed
huge swaths of Indian grass, wild rye, thistle.
Mice, rabbits darted, pheasants shot up
as my beastly John Deere cast its shadow
and shuddered along runways, cut a six-foot wide
strip by quarter-mile-long strip seven hours a day.

Those were the best Cokes, eight ounce bottles
the gang boss ferried out at around 2 p. m.
in a beat up Ford F-100 painted like a yam.
Iced, shot-cold bottles tipped upside down.
Run fast down a throat scratchy as hay
under a sun that tried to teach me what it takes
to be relentless, naked, generative without pause.

Come Saturday Clare and I, bored out of our ever-loving minds,
would drive to auto lots along Ogden Avenue.
Pretend we were in the market for this GTO,
that crouching red Cougar. Fudds in starched shirts
and a gas cloud of spray-on deodorant would
have to stuff their *Playboys* into side drawers,
swing loafers down off the desk, toss on a lime and pumpkin
plaid sports coat. Then hoof it out into summer heat.
After we gave them our licenses to hold,
and fully against their better judgement,
they grudgingly forked over keys to spanking new machines.

Off we'd spin. Crank that radio. Find the Stones or 'Retha on there.
"Goose it Clare! Give it something. See what's under the hood."
Back then punks were made to go places. It was one way
of telling ourselves we weren't going to spend the rest
of our lives tasting farm dust. So we flamed out
large mazes around these lots. Circled them like we were
buzzards floating high on top of fabulous fountains of wind
over an endless range of mountains called future, called fate.

Too Much and Not Enough

Sherry had a laugh that brooked quickly
through the aisles of our drugstore.
Everyone stopped to listen,
see if it would come again.
She had high points, blouses tucked smartly into skirts.
Was married but... As she tended the cash register,
Sherry had a way of letting her hips settle
into pure, restful satisfaction.

Smart too. She lived in a two-story clapboard
on Klieg St. Winter mornings I'd chug by
and see her DeSoto Firedome buried in a cave of exhaust
while she was still toasty inside doing up her lashes
as the slant six warmed with a brick propped on the gas.

I didn't have a clue why our eldest cousin, Clint,
dropped out of grad school at Marquette. Came back
to work under weak fluorescence for Dad.
Then left abruptly to become a Zenith sales rep.
Stopped eating. Got bags dark as sewage under his eyes.
It was Sherry's laughter. It was Sherry's hips.
It was what flashes of heaven a man can almost get to
in a suite at the Holiday Inn or in a DeSoto Firedome.

I learned that fact years later when Clint went psycho.
Pulled a kitchen knife out of the drawer on our grandmother.
Then ran cursing out into winter. Drove down Klieg Street.
Luckily Sherry and her husband's door was bolted.
Clint stood there, shouted her name. Ranted up at
ice-cased windows. In a nightgown, Sherry jerked
an upstairs window open, yelled, "Go away!"
Screamed, "Just leave us alone!"

Finally the cops came to haul Clint in. Cuffed him
though they didn't know what to do
with a three-sport high school star who had gone
stark, raving lunatic. This got passed over
in "The Weekly Gab-About" whose front page that week
had a photo of a local bowling team
that had punched their ticket for Nationals in Tulsa, Okla.
where they might finally show everybody how
ours was a town that knew how to set things up,
knock 'em down. Set things up, knock 'em down.

White Knight
Talking Backwards

In the Summer of Love, I was seventeen
when my sister let me out beside the Interstate
in Plainfield, IL. with seventeen dollars and a thumb.

In the Summer of Love a curly-headed
drop out from K. U., Lawrence told me
about the time he licked an acid tab
and drove home watching wee people play
poker on the hood of his VW bug.

In the Summer of Love, I slept in sand and
pebbles on a ridge not far off an Albuquerque exit.
I listened into a star-spackled dark where
coyotes and wolves yipped, howled.
I woke up shivering. In the Sierra Nevada
I slipped into the back seat of a Dodge
parked for repairs outside a mechanic's shop.

In the Summer of Love, a biker put his tongue
down the throat of some long legged thing
in the doorway to HIS PLACE
where everybody on "The Strip" hung out
because they brought quartered peanut butter sandwiches
in on paper plates every half hour as a way
of introducing us to the sacred heart of Jesus.

In the Summer of Love, I met people who claimed
to be Jim Morrison's only brother on three separate occasions
though they never looked remotely the same except
for those trademark cascading ringlets.

It was after midnight on the Sunset Strip
in the Summer of Love when I put a calming arm
around some half-pint, acid-head from who-knows-where
who thought roaches were crawling over his skin.
The next day a head-shaved Moony in a van told me
he was learning to be one with everything then
gestured to a patch between I-5 and an entrance
ramp where a lone sapling stood, leafless.

In the Summer of Love, I left LA. Listened
deeply as a man whistled a three octave aria
and his Impala threaded through warp
and weft of mountain valley, mountain crest.
I caught a hitch from a long-hair in a pickup
that swung us by his house in the canyon where
I shared tea on a mattress in the corner
with a girl named Alice. No more than fifteen,
Alice had run away to this house
where she swept and washed all week
and the wife doled out LSD on Sundays.

I tried to touch her with my voice, but had
no idea that I was still thousands of miles
and over two years away from love.

HELL OF A RIDE

One afternoon on Route 66 I rode
with a laundry bag down between my legs.
It contained: The Complete Works of William Blake,
four tee-shirts, an extra pair of jeans, white underwear,
socks, a loose-leaf notebook, a rolled blanket.

I was in the shotgun seat of a death-black
'51 Chevy rolling west. In diapers, on fat legs,
a babe bounced up and down on the seat
between me and his dad who gripped the wheel
with one paw, had a Bud in the other. The babe
shook his hands up and down with glee.

Light was sloshing over every green toss of branch,
bolt of cut creek, pasture, field. All lush,
as his souped-up Bel Air rocketed up to ninety eight.

The driver told me how he'd ginned that engine up:
headers, dual-point distributors, chrome carburetor.
There wasn't "a cop in four counties" that could catch him.

When we bumped up over a crest, the guy's
right arm, the one with the Bud can in it,
shot out to steady the toddler who had
lurched forward but was then made fine.
As a splat of suds sloshed out onto seat fabric
the driver let out one quick-bitten curse
that was more like an appreciation
of how good that beer was to him right then.

We passed a sign that said "Poplar, Missouri,
Population 501" and I thought, probably 502 now
with the addition of this bright babbler at my side
if he lives long enough to be in the next census.
The Chevy stopped to let me out, turned slowly
down a long, shaded lane. My parents, brothers, sisters,
the high school I had just escaped were two states behind.

That night I ended up sleeping in Joplin, tucked along
the side of a garage within sight of an A & W.
I woke around four paralyzed by a dream
in which three Doberman Pinschers leapt off
the garage roof but were suspended above me
in a sort of mobile of terror. I got up,
walked out of town west, towards California.

IN '69

I hitchhiked Route Six Six, Chi-town to LA.
Took four days but there I sat next to an off duty cop
in a Buick '98er at ten p.m. while cars whizzed quick
and orderly as electrons around so many loops and freeways
I would not have known where to begin.

So this kindly soul drove me all the way down
to the famous Sunset Strip where I'd seen
Edd Kookie Byrnes flip back his 'do, hop into a red Corvette.
I wasn't on the street twenty minutes when some Jack
in a turtle neck asked real nice if I'd care for a Coke
at Vipps, Home of the Big Boy. Green, I said, "Sure,"
a little perplexed until he brought up his proposition.

I slept in a back alley where other hippies before me
had flopped down mattresses. LA had air just that calm
where you could lay out and sleep under blotchy stars
with barely a stitch of blanket. A dozen of us there. Some
shared left-handed cigarettes. Nobody one whit Californian.

The next night I tried to find shelter in a missionary quarters
run by a Born-again outfit in a loft where they set out
mounds of peanut butter and jelly sandwiches every half hour.
The loft was packed with grungy types. When I asked
someone how to find a job, he said. "Do like everybody else.
Get a hotel room, stick your thumb out. You'll make enough."

I sat on one of two couches which were the only furniture
in all three rooms except for the sandwich table
and a twelve foot cross that jutted up into a skylight.
I was waiting for my interview with one of the ministers
who also offered free bunks in the back rooms, safety.
Some teeny-bopper slipped into a sneeze of silk blouse
and down from the hills slumming plopped down
next to me and apropos of nothing said,
"I want to rip something off." I had never heard the phrase.
I hoped she meant her blouse. To this day,
I can not imagine what she was thinking.

I stayed three miserable days in L A. Never did get a job.
Did not even make it to the Pacific. Met too many
blown to nowhere fools. Never even saw much worth stealing.
Lit out of there for home where I settled back in like
yet another Dorothy grateful to be back in their own Kansas.

INSATIABLE
THROAT

When I First Heard "Black Magic Woman"

I remember that guitar

                    curled,

                          peeled off ribbons

that ended

           somewhere above

                  and behind

as we flew up I-94.

Gulped wind,

            headlong drunk,

keen-eyed.

      The hearts of wolves

             in bone cages.

I remember the size

            of darkness.

The fantastic

          scroll and flow

             of bent steel

          that cried out,

                with elegant intent

as it escaped from the radio
                     of a Dodge
                          that didn't
         much care
                where it was headed

over pavement poured
                     from one side of the country
                        to the other,
under street lamps
           that can't even begin to expose
                   how much hunger and ache
any one night contains.

WHEELS OF DESIRE

I had a beaut of a '60 Chrysler my second year
in Wisconsin at a college perched atop
a limestone ledge beside Lake Michigan
which seethed and sawed at its horizon.

Pushing buttons on the car's dash,
I'd jolt from Park to Neutral, Neutral
to Drive. Then stomp that pedal down.
Make Goodyears eat the highway up.

My monstrous Chrysler was bronze.
Had fins fit for a shark at the back.
The speedometer was a red bar that raced
to the right and lay inside a ginormous bubble
round as Alan Shepard's helmet which put
his head in a fish bowl as he golfed on lunar dust.

Mostly the Chrysler collected what ice
and snow the winds off Lake Michigan delivered
for months on end. Dried-up, curled leaves
lay frozen in the wipers' gutter; I'd curse
and jab a scrapper down along windshield glass.
Try to turn that bastard over and keep it going
so I could get to a two-bit drugstore job.

Those were strange times. The night was
bigger because I had so little reference on it.
Stroking a co-ed was pure electricity. I wasn't
always sure why I studied what I studied.
But come Saturday I had a bronze Chrysler
ready for blast off. Fish fins with red taillights
sliced through swirls of snow as we raced
into the headlands of desire, which might
have just meant picking up pepperoni pizza,
as after all, I was in Wisconsin.

I Hitchhiked All Over

Wheat weighted with snow.
Bone moon working like a bleach.
Corn stubble, stark, vast.

Weeds up through a rusting Chevy
whose hood yawned for no reason
other than laziness in country shade.
An enamel freezer on a porch
beside a seat unbolted from a Plymouth,
then set as an outside couch where a man
could pass the night, smoke, keen for owls,
piss beer just beyond pine boards.

Riverbed swaddled in cottonwood.
Arroyo run down from mountain snow.
That mountain turning in the distance
somewhere between blue and purple
as the sun fizzles on the horizon.

Woman in a lipstick red dress with
huge polka dots, hose and high heals
stuffing notes into her purse
as she steps out of the State Bank.

It all had the same weight to me.
I needed it bad. All of it.
Had to get to everywhere fast.

I stuck out my thumb. Waited.
Some goober jerked his pickup
to the side. I ran to catch him.
"Where you going?" he'd ask.

I'd talk up a trucker by a gas pump.
He'd say, "I can get ya
far as the Oklahoma line for sure.
You're gunna have to crouch way down
outta sight if I spy a company man.
Where ya headed anyhow?"

I always told 'em,
"Up the road a-ways."

MENOMONEE ET AL

For truckers hauling threshers and reapers,
Wisconsin was a pinball map lit
up with Indian names. Waukesha,
Wauwatosa, Menomonee Falls, Oshkosh.

High school grads hunkered down in apartments or in basements
under the shifting floorboards of their mom and dad
without much more than the tube, a fridge, beer.
Winter pipes ticked, cracked. Engine blocks got so cold
they needed an electric blanket of their own.
Rivers froze except in town parks where
falls still thundered as a sweatered dachshund
lifted a leg to milkweed, sizzled snow.

Every night glacial clouds drifted.
Winter sky went so black eighteen wheelers
slid off highways, had to be dug out of drifts.

It was a state under siege come January.
Way more gravestones than there were tourists.
So much reticence in its Germans, Slavs, Poles
that you sensed the state's Indian heritage behind
each begrudging movement as these
pseudo-Indians sat the couch, downed their Blatz.
There were no braves left to straddle a spotted pony.
No proud profile to gaze out atop a limestone cliff.

At best, some teens got loose. Hopped up on hootch,
they'd whoop and guide an F-10 Series Ford
down the Wisconsin River's iced and tortured curves.

On the Loose

for Gary Beaumier

A friend and I hitched to Mad-town in a blizzard in January of '70
before Nixon got impeached – around the time he cooked up
a "Draft Lottery" which gave Gary a bull's-eye on his back
when his birthday was pulled at number seven.
Sent me skipping free with number two-one-three.

We shivered in the back of a sled – Monte Carlo –
whose heater didn't have much left to give.
Listened to cousins who handed back cans of Schlitz
from six packs they kept at their feet under a towel.
The interstate was well salted; single file where a dozer blade
had passed, packed a running C-shaped wall to the right.
Fat flakes in constellations on the windshield
melted by the time the wiper slid up,
got shot to diamond in the rare moments when
eastbound headlights speared into the cab.

These cousins had picked us up outside of Waukesha
where we shivered under hoods. Hopped side to side.
Listened to the crunch, peculiar whine this sort of snow
makes under boots. No holding us back. There were
co-eds in Mad-town we just knew wanted to meet
two farm town yokels who could recite all of Dylan's,
all of Simon and Garfunkel's, lyrics like they were poems.

Anyway, these guys were lit and talking it up.
One said "Remember that time I picked Stan up
from the state house? We went to that dive in Saint Paul.
It was freezing like this and everybody got totally shit-faced
tossing back boiler-makers 'til Glen sailed off into
sting-ass cold to start doing somersaults and handstands
up and down Main Street under the damn street lights.
We all did 'til finally, and sure enough, the law showed up.
So we went from drunk and disorderly to drunk
and orderly in the flash of a lighter, just to keep
Stan from going straight back to the slammer
in Faribault. Now that was fucking living."

The cousins set us down in front of the Rathskeller
which was covered in snow but still glowing
like a flashlight under a blanket with a kid up way
past lights-out reading fast from a book he just knows
contains everything he can ever imagine and then some.

Man Sans Car

In 1971 I biked to the jetty at red sunset to watch
fish boats nodding home while gulls circled their stacks.
Watched thin sheens of petrol shimmer
over rollers as Lake Michigan lashed at my feet.

Biked to antique stores to start a collection of 78s.
Always toward the back. Always leaning
heavily one upon the other in tattered cardboard boxes
because no one would love them anymore but me.
> *"Eyes that glow with the darkest light*
> *I'm going to serenade you tonight"*
Up by the iced window under the arms of an Aurelia
the store owner jotted in a book within slant sunset light
as a Tiffany lamp added its cast of prisms.

All because I went Zen in Racine, Wisconsin.
Gave away my car, almost everything. Kept
a Bartok piano suite, the Ninth, my Blake and
a well-thumbed copy of "Riprap" by Gary Snyder.
If you're sitting in lotus, you don't need a couch.
I learned exacting rituals of leaf bits and twigs –
the methods venerable monks use to brew tea in Japan.
That they splash in a ladleful of brook water at the end.

I got around on a red J. C. Higgins I'd been
gifted under the Christmas tree at age twelve.
Instead of buying a lock, I wrenched the seat off.
Rode standing up. No one ever tried to steal it.
For gloves, I wore socks. Rode everywhere.

Kicked the bike aside and lay down at a curb
to watch the city take place from a cat's vantage,
its parked cars, the asbestos-slatted sides of houses.
A woman burst out, whipped her broom over the threshold
and across a cement landing. She glared at me with furious disdain,
adjusted her bandana, bustled back inside.

I rode down side streets between brick and mortar
then stopped in dead winter to hoist myself up, peer
into a cavernous foundry, see its charred ceiling,
steel catwalks, hard hats, vats pouring furious orange
metals down into lines of ingots that glowed sun-wild.
I rode and rode, was restless as the lake.
Only still when I went home. I was an ameba with
ill-defined edges that engulfed everything within range.

Apropos of What the Hell

for Jack Albert

Jack came from Philly, had a leather jacket
with sleeves full of sly. Leaned in a way,
hips almost winking. Everything about
him sideways and ready for the night.

Gamblers, story-tellers, dark-eyed women,
always found him. You might be
a professor, be the youngest daughter
of a Proctor and Gamble's senior vice president
– you'd get drawn in. He had that way.

Jack was a welter weight, golden-glove, one
night. Drove vans for the mafia the next.
Said that back in Philly he'd run with hoods
who lurked *outside* the stage door of *American Bandstand*.
Roughed up the pretty boys as they came out. Stole their girls.

I joined in once, hitching south from
Racine to catch a Chicago transit.
We got picked up by this thin spoon
everything was wrong with.

He had grease licks in his hair.
Looked like he'd just escaped a bowling alley
but was also stuffed into a sports coat.
Drove a Maverick, was headed toward a sales rep meeting
but only had a titmouse's sort of confidence.

Jack cocked back. Said we were on our way to Cicero
to shoot sex scenes with five coke-dusted slut-lets.
Things lean, curvaceous, pink-tinged, frothy.
That we did this couple-a-three times a month.

That it paid in Bens. This sad stooge got so fused
he jerked the car to the curb, pounded the steering wheel.
Cursed, "And here I am, married
to my hand in Kenosha, Wisconsin."

INSATIABLE THROAT

midnight   Turnpike

Allegheny mountains   obsidian

ruby taillights duck behind bends

other riders   hunker silent   under coats

a lone trucker sizzles past   flicks

a cigarette butt   that hits

        bounds
        comet tail

sparks backwards

into rearview   scuttles

down in dark   under stars

I left   a city   and a woman

        nine hours ago

The insatiable   throat of the highway

swallows me whole

Going Through the Gears

In a Gravel Lot

Charlie had a Ford Bronco, had a muck-sloshing brush-crasher,
back before anybody had that shit. Knew what to do with it.
In '72 he shagged up to Vermont for a writers' conference.
Less than forty-eight hours later he had us wheeling
through the Green Mountains, whipping around hairpin turns.
He screeched to a halt – cupped hands to catch trickles
as they trailed off shale and moss down a cliff wall.

It was Sunday afternoon in back country. We were out
for a clogging fest that we found on a lawn beside a Baptist church.
Every Joe and Sadie there except us seemed at least
distant relations and had traditions that harkened back
to when Mohawk warriors showed up at trading posts.
Folks still wore overalls, grinned tooth-poor as hockey pros,
taught us steps, played reels. Wore the day out pouring jug whiskey
down our throats until I finally ended up shit-faced and slumped out
in the back seat watching the Green Mountains sweep by
with Venus set inside a prism of sunset in the side view mirror.

A couple years later Charlie and I mushed that sled
from Louisiana to Wisconsin and back on a three day weekend.
I showed him my town, its golden wheat, river rife with muskrat.
Took him to the dim grocery, all six of our gas pumps,
the dim bar, the dim bowling alley and our town library
open between 10 and 3 Tuesdays, Thursdays, Saturdays.

On the way back, since Dad knew we were planning
to zip straight down I-55, my pharmacist father
slipped us a couple of bennies.
That's how truckers made it back then.
So about an hour past Memphis with the sun
cropping tree tops and Charlie burrowed deep
into the crack between window glass and shotgun,
I spy a sign for a greasy spoon, jostle Charlie awake.

He squints as we get out onto the gravel lot.
Me, I'm something else. Something I've never been before.
Never been since. We walk slow, steady. I talk
in just above a whisper. The whole scene's gone feathery,
like how it had to be for Jove pillowed aloft on clouds,
except more so with every bone somehow pliant.
I tell Charlie, "I feel like I'm James Dean." He looks at me
and says, "You are James Dean, you are James f-ing Dean."
Heart pumping God juice, we uncork ourselves from a Ford Bronco
sunken deep in northwest Boondoggle, Mississippi.

Sacrilege and Sacrament

John had a beat to hell white pickup
he inherited from an uncle and brought to LSU
where he practiced the art of nihilism.
He said all he did back then was button his lip.
Try to breathe in what smoke and light he could
from the night fires of Friedrich Nietzsche.
Which factor, along with his thin wallet,
might have sent his first wife packing
to be alone in Tampa Bay where she didn't
have to worry beside a petulant man.

That was before I met John. Before he lifted himself
out of a philosophic stupor to teach me how to drink
like a southerner and how to navigate a two-step.
That was well ahead of the time I am writing about
which is the night before John's second marriage.

It started out as a bachelor party. A slew of us
and copious amounts of bourbon were together
on a front porch which had already been the scene
of so many late nights that a neighbor later told us
they started a novel to chronicle what went on there.

What they imagined taking place a few steps off the porch in
shadows given to secrets between men, between men and women,
between women and the night. These were good times.
These were the very good times hungry, half-formed people find
to gorge themselves on so they have thunderous memories
sixty years later in quiet rockers on porches.

A flock of us kicked it off drinking the way peckerwoods will
when they start an evening trying to carry sun's radiance
along with them into pitch. Then come to twelve hours later
in a wrestle-hold with blind darkness in broad daylight.

Charlie De was there whipping us to a frenzy.
His twin, Johnny De stayed to the side, the way
a helmsman will – intent always on the compass,
the wake, the pitch of the ship and what path he sets.
I remember that Ben Gold was there too, because
it was a year before he became possessed,
left his Delta Delta Delta wife. Flew off
to "Zion" to fire an M10 into desert sand
up and down the borders of the Holy Land.

Others, too but they thinned out. Sauntered back
to their homes until it was just the five of us.
We decided to do what we did back then
which was to jump our drunk butts into John's pickup.
Head out to the levy road because of how it sweeps
beside the Mississippi River the way a violin
swoops and spools in consort with a tenor's lead.

We stopped short. Stopped beside a field.
I have no memory of why. Maybe because it looked
like there was enough room, enough isolation.
I don't know. What makes a dog pick one patch
of weed to pee on, not another?

We were there. Stars were limitless.
The field showed fresh tilled rows.
Maybe that was it, a promise the field gave.
Like we could plant it. Like it would
fruit for us and for John and Becky's future.

So we stumbled around on packed shell
at the side of a forgettable highway.
We whizzed in the weeds. We sang.
We lay across the hood, engine heat to our backs.

The five of us hooted, hollered, banged
the hollowed out sides of that beat up truck
as a trillion stars drove godzillions of crickets
through relentless hosannas. We acted
like we were sending our brother off to die
in the rice paddies of Southeast Asia
instead of just off to get smothered
by the holy sacrament of matrimony.

CHEVY PICKUP, LOADED

for Nicholas Zakis

Back while he lived on the Bayou, worked as a derrick hand,
Nick had a Chevy Silverado, clean as a sword.
When I hitched down there to find temp work
during breaks at LSU we'd spin out
to the Hubba-Hubba Lounge where the juke box
always twanged out merciless, teardrop crap.

Roughnecks just in from a two-week shift could be seen
dancing with their cousin's wife at two in the afternoon.
That cousin now out for his own two-week stint on a drilling floor
as this little Cajun queen lived it up for all she was worth
at the Hubba-Hubba Lounge mid-afternoon in Galliano, LA.

We just sipped suds until the phone rang
and the bar maid picked up. Then turned and asked,
"Anybody want a deck-hand job with Nolte Theriot?"
Two hours later I'd be in a company van
headed to Sabine Pass, state of Texas.

Nick kept an English gent's black umbrella on the rack
in the rear cab window as a "fuck-you-very-much"
to the rednecks that all set Winchesters there.
We'd wheel along the bayou counting gator heads,
flipping empties out the windows to let wind
sweep them back into the pickup's bed.

When he made the long trek to Baton Rouge,
Nick parked behind my rent shack so the repo man couldn't spy
his Silverado sitting hot, ticking down under the trill of crickets.
He'd grab his Gibson, pick out Cohen songs on the front porch
or we'd hit the bars and disco clubs along College Drive
where Nick danced like an electrified rooster.

There was no stopping that boy back then.
No trouble too big anywhere he went.
That Silverado must have finally got paid off
but it never did – no how, no way – run
down one single road headed into tomorrow.

WHERE JERRY TOOK ME

I found "Jerry" the big Ford Galaxy
under gum trees in an old man's driveway,
tire rubber crumbling into concrete.

I coughed up two-hundred-twenty in U. S. tender
for the right to splash gallons of petrol
into a tank large as a cow's belly.
For a way to throttle out into barroom nights
where torsos swayed to the beat of drums.

One woman had left me for her husband.
I had left one because I would not be her answer.
I was unhinged. Now I had a vehicle for that.

Jerry nosed along the levee, sailed the interstate.
Jerry's floor wells accepted flimsy bullet-smooth,
aluminum cans as we quaffed effervescent, golden liquid.
Jerry soared me over asphalt toward what oblivion

I could find in the folds of any woman
I convinced I could do right for a few hours.
I remember storms of sheets, crooks of elbows,
eyes that blinked back open. Shadows
of socks, bra, pants settled in quiet pools.

I remember a splay of hair backlit by dawn window
where a trellis of heartleaf philodendron
tumbled behind bare shoulders.
Beyond it all, the belling of hips
that contain what night contains.

THE ROAD TO ALLIGATOR BAYOU

Nick and I found a gravel road,
with cow fields to one side,
Bayou Manchac on the other.
Midway he stopped the pickup. Flew out.
Crouched. Clicked a photo of an iridescent snake,
lime, thin, blended into bamboo stalks.

I kept going back for decades.
Around one bend was the Alligator Bayou Bar.
On a pointless afternoon nobody cared
that its plywood floor drifted
crazy as a potato chip and made
the cue ball dip as it tottered into inertia.
Saturdays when some Coonass band
went at it hard, men left Buds on the bar
to whirl and spin with their women.

I'd shut my headlights down on full moon nights.
Pour past sleeping cows under blotches of shadow
and silver, road curves banked like a river.
One April night, katydids, crickets, tree frogs
all in full throat, a nurse got out of her skirt
for all that trembling beauty. I lay my jacket
down in the middle of the road. We became
one pool together welling within
the pool of moonlight welling.

I Want To Get To Havana Before It's Too Late

When I went to Lima in 1979
I often rode in "colectivos" that tooled
their four-mile stretch of boulevard
all day long. Every few blocks they picked up
sisters out to shop, a gringo, or a pin-striped lawyer.

The driver let us out after we tapped his shoulder,
whistled or called. We crawled out over toes and knees
as if we were mid-row at a baseball game.
It was a warm environment where any body
lounged, laughed or snuggled, spaced wide
or pelvic bone to pelvic bone.

Colectivos were easy to hail.
A black '48 Chevy with hip-round fenders
and an oval rear window fit for an admiral.
A '61 Chrysler all chrome, glam, angles.
Every one of them plush and roomy. Still elegant.

The women without babies inhaled slender cigarettes
deeply, with nonchalant intent. Those with children
kept them in, chicks under wing,
as buildings, street signs whirred past.

On one corner I saw a sign for a mechanic's shop
that said, "Nos especializamos en Packards & Nashes."

BAD WORK

Sometime in the Seventies
I saw Volvos start to pop up
at city stop lights
with drop-dead beauties
framed in window glass.

Then I danced with one
dressed as a de Kooning
at a party thrown by an art professor.
Paula. Paula could shimmy,
could hesitate fragile as ash
for one drum beat
at the end of my grasp
then sail in, swirl under
my uplifted arm.

Soon we started to drive out
to a bar pit by the interstate.
We'd lurch toward that lake
through a quarter mile of twists,
crusted ruts, slop. Volvo headlights
lit on armadillos under shrubbery.

We'd smoke some weed, down wine.
Lay out suspended, bare,
in star-touched waters
as tree tops rimmed our vision.

Eighteen wheelers barreled
through distant interstate pitch.
Relieved of gravity, we drifted
like compass needles held fast
by the true North of a magnetism
that arises in the bodies of lovers.

A few months later, living together
in the sacrament of our hope,
her Volvo's brakes started to squeal.
I said we could save a bundle
if I replaced the pads myself.
I crouched, knelt, grunted,
lay on my back.
I threw all I had
into splitting asunder
nuts fused to bolts
by thousands of miles
of furious motion.
Down driveway concrete
I laid parts out in radiant rows.
She fixed a sandwich with chips.
Brought me ice water.

With gritty determination,
I wrenched it back together
but didn't understand a thing
about fluid or pressure.
So the next day
Paula flew through the stop sign
at the end of our street
into a vacant lot big enough
that she could run the Volvo
down in circles.

The crazy thing was
that she forgave me.
Later that year we married.
For various reasons
more complicated
than any mechanical work,
I mucked that up, too.
Before we separated
the Volvo's engine froze,
so the car got replaced
by the most practical, boring tin can
this skinflint could find
in the Want Ads.

Our divorce gave me that car.
I did nothing but go to work
and back for a long time.
I eased my foot off the gas,
hugged my lane.

THE BROWNED BUD
"She was a long, cool woman in a black dress." – the Hollies

Though we quickly forgot much of the poetry read,
many of us couldn't get past how she crossed,
then uncrossed her legs, straightened her back
seated at a table just off the poet's podium while
Naomi Shihab Nye, Carolyn Forche, two others read.

At the after party professors, grad students
vied to bring her drinks, did buck dances before her.
When my friend Mike said he had to go home,
she was at the kitchen counter near my keys.
I asked if she wanted to catch some air.
"Sure, why not." Her whole frame sighed right.

Nothing happened till after my headlights shot up against
Mike's garage door. After I let him out so he could get back
to his good wife. Before we even made four blocks
my fingers broke through a gap in her fishnet stockings.

We drove to the lake where the moon came strong
across a chop of waves. We swept down onto lawn
close up to a Camellia bush whose heavy buds
exploded over her shoulder, lay littered in the grass.

After it all, still lying there, she brushed a browned
flower-head away from where it was touching her wrist.
Then thought better of it, picked the bloom up.
Lay it down on her belly in moonlight.

We went back to the party. She told me her name
was "Fawn" before zooming off in an Alfa Romeo.
I drove my Toyota back to my sleeping wife.
A year later, as part of the mess of our marriage unraveling,
I ended up confessing this. My wife was crushed, tormented.

Even if all I'd ever been was a gangly mess from a farm town
who often made unfortunate sock choices
and no matter how hard the stars bare down
on any of us who live for love, I knew better.

Wild, Wild Horses

Afterwards I moved to a duplex apartment
in a quaint, cat-friendly part of the city.
I stayed quiet, redid a common courtyard
the absentee landlord couldn't have cared about less.
Put in two cypresses, sweet-olive, a flower bed.

Technically she was still my wife when an acquaintance –
an anorexic painter who started out as her friend –
confided that Paula had that very day gone
to Jazz Fest with some bass man in a punk band.

At around nine that night I plucked a knife
out of a kitchen drawer and drove to the old house.
Parked at the curb under street lights' iron shadows.

It is lucky, for sure, that the two didn't show up –
that bass man and my somewhat wife.
Maybe stopped at a bar to drop Ecstasy, toss a few back.

Lucky that I never got out of my sad Toyota.
Drove off after rage started to embarrass itself
while the Stones blared dead on into night.

I've Had a Bunch of Beaters

It was a mechanically sound Corona station wagon,
boring as a night spent with a visiting aunt.
You could soft boil an egg in the time it took
to get this thing up to seventy, feel them struts shimmy.

Ran it into the dirt going up and down an interstate corridor
for one of those jobs where you watch the dead
come alive as they rush out the door every 4:30 pm
toward cars parked in the punishing brilliance of a shell lot.
Building cleared by 4:33. Last out trips the alarm.

I used to take it down along River Road,
boom out, swerve back like I was following
the lazy loop of a saxophone's neck.
Discovered a church in Bayou Goula not much bigger
than a walk in closet. You let yourself in with a key
kept in a small cypress box off to the side.
Inside, to the left, was a stand of candles whose flames willow
when the door opens or closes. They left a butane lighter
and a long stick you snuff in sand after giving an offering.
That way your prayer stays behind to cast jasmine light
onto a Saint Christopher, a Francis of Assisi, four distinct Marys,
one with a serpent crushed beneath her left foot. That is the way
Louisiana is – licked by rivers, flush with faith.
A place where you hear songs well up even in its silences.

I finally dealt that monkey-shit brown Corona
to a little guy from Gonzales on a Sunday
when nothing was open to do the paper work.
He promised to register the sale straight away, but evidently
never got around to that because a year later I got a call
from a detective down in New Orleans who wanted to know
if I still owned a brown Toyota station wagon. Because,
in point of fact, it was a subject of criminal inquiry.

I laughed out loud when he said my own special beater
had been used in a snatch-n-go robbery of some poor bastard's
newsstand at the corner of Hopeless and Gentilly.
I couldn't imagine the getaway scene as anything other than
a slapstick Western where harebrained cowpokes
hold up a noon coach. Then try to make a getaway
furiously switching the flanks of burros.
I envisioned a trio of sad sacks back
At their apartment counting the haul,
carefully calculating how many more heists
it was going to take before they got enough together
for a vehicle worth half-a-damn.

On the Right Exactly Seven Tenths of a Mile Past Eureka Springs Free Will Baptist Church

Disturbed as my house cat had once been on an afternoon
it spent being dive bombed by blue jays for hours,
I drive my shitball Toyota station wagon
between horrific lightning strikes in the Ozarks.

When I finally get to Jack's dirt-pack driveway,
sky clears, opens like a corn flower.
Jack is a Philly-born college classmate who grew up
to sell off his small Manhattan restaurant and move to a holler
in Arkansas with his dancer bride, both suddenly thrilled
to be a thousand miles from N.Y. under hordes of stars.

I get to their campsite of a home before they are off work
to find a four-foot rattler sunning itself on flat stones
that Jack had hauled out of the nearby brook
to form a fire circle just a dozen steps from their tent.

Seems the snake feels he and his were there first
and, "Thank you very much for these fine rocks."
Unnerves me enough that I hunt up a shovel,
drive its blade straight through the serpent's spine.

Jack and Cindy can't thank me enough. Have been
scared of what karma they'd get by offing a sentient being.
Which is not problematic for yours truly, as I figure
I've already accrued the karma of a poisonous sidewinder.
We jump in the creek. Jump out. Light up a ganja cigar.
Lay back to watch how cliff sides work their way through eons.

The next morning as we drive out through a field of weeds
Jack says, "Wait. I see Hermit George. He's pretty funny."
Sometimes you worry your bucket of bolts
won't be able to get you up mountain roads.
Sometimes it takes you too far, because
sure-as-shootin' once grizzled, young Hermit George
sidles up to our door-side to grin his toothy grin,

Jack asks George what he's doing this fine morning.
George says plain as newsprint, "Eatin' ticks."
Then does a little jig. Crickets all he has for fiddlers.

TIME AND
THE WHEEL

Into the Blue

Had a powder blue Mazda,
a chunk of sky on wheels
that slow-dripped oil, sank
a two-d cloud into concrete.

Had noon sky on wheels
that set a dark slick inside a shade skirt
under powder blue in my car port.

Had two children that rushed over grass in jellies
drove themselves forward and back to wrack and work
the springs on an otherwise inert hobbyhorse
in the corner of a room off the kitchen.

Children squeal, pout, crouch, wail, twirl,
dash, swing, whinny, jump but mine also –
if I said we were going to so and so
for such and such in about a half hour –
dropped it all. Scampered off to get a book,
grab and drag a big doll by the leg.

Put that stuff up on the unflappable firmness
of our Mazda's back seat. Hauled their little selves in
first pulling with their arms. Then they got their bellies
onto the seat, 'gatored on up, dimpled thigh-backs, et al.
Arranged themselves in ersatz comfort.

Looked at pictures. Pretend-read what I'd
read to them the night before. Even at four
my eldest once said – turning back a page to check –
"Yeah, that makes sense – It's the same morning
but now she (Amelia Bedelia) has a red dress on."

That is where they'd be when I finally got my coffee
into its go-mug, locked the backdoor. I'd find them,
hair matted in sweat in the swelter-crush of summer Louisiana.
Even that couldn't wholly cook out their wild anticipation
for what any day might hold once Daddy
put his blue cloud into gear; let it roll. Let it roll.

When Cars Were America

Granny got a Caddie.
Got a hold of gold keys given
by her doctor son one Christmas.

She said, "Pile in my polliwog.
Pile in the back seat."
She'd fly off to Seven-Eleven
for Icees after pre-school.
Big air-puffed Coke, strawberry Coke,
you suck down with a fat straw.
Gives you shivers in your throat-back
in the deep comfort of "Granny's" Cadillac.

Blue as the big, blue sea, Granny's
Caddie whizzed along the fast lanes.
It had photosensitivity too;
a bitty-chip that shut the
lights down. Shut the lights down
when stubby-as-a-gremlin Granny
didn't remember to do that thing.

How she grinned to tell her cousins
that the M.D. son gave his mum
her big, blue Caddie Christmas.
Knew she knew they knew
she's the very one that made it.
Became the apple in the family pie.
Then went zoom-zoom on and on.

A Right Then a Left

for Miriam Ruzicka

As we razored down an Alpine highway,
gas gargled in the Peugeot Junior's carburetor.
But you, tiny cricket, tid-bit, oh no! Were you hungry?
I'd checked; you didn't need a change.

Cranky for no reason, you screamed
intolerably long. Loud, little smidgeon,
you of the gifted lungs. When I took
a road to the right, we swooped down
from high bright air to sink under the shoulders,
flanks, thighs of a towering forest.

So sudden, this forest shade, that you
gasped in a doll's cup of breath once and again.
Inside the valley's branches, crags and
rock, dark was wrestling gold while
gold was wrestling green. Sobs stopped.

Within the deepening chasm of the Loue valley,
breaths came cool and cool again.
We ran beside a trout-quick river,
its rills still shot silver with a tumbling coinage.

After we ate at a restaurant, you dozed
as the moon sent down great shafts.
Now onto water-batted, stream-dividing boulders.
Now to a birdbath, fence and flower row.
Another hour at that wheel. Ninety degree
turn. Motel door flooded bright by headlamps.

Time and the Wheel

At a stoplight on Airline, headed home,
I had a twenty-minute drive left
after my day's final treatment
in which a patient who had lost
forty pounds and everything
after a West Nile mosquito bite
was able to go from sit to stand by himself
for the first time in four months.

I looked without looking at pavement
in a Quick Stop parking lot.
My eyes followed the calf lines
of a mother who gassed up as her children
bickered in backseat heat.
I saw how expertly she pivoted
the hose handle back in its holster.
Punched a gas-pump keyboard twice.

A man rode up under a gold hat
on a bike spray painted gold,
gold sequins sewn onto pants.
Right after he sat down on a curb
beside an older man who had a bottle in a brown bag,
the older man passed golden beer
to the man decked in gold.

I took in the low angle of sun,
how it skinned car metal in a kitty-corner lane
and came up bright in a puddle beside a Dempsy Dumpster
not far from where the two men had their shoes.

I sat there till the light finally greened
and thought, "This is nice. So nice to just
get a chance to relax here where life sheds
what it sheds every second, to kick back
and breathe without another morsel on my plate."
That is when I realized that I had been
working way too much.

What a Pass Key Gives

Hotel rooms come set
for luxurious sleep:
safe, impeccable

with deeply creased sheets,
heavy curtains to block
the least shred of light.

We drink beer, make love,
discard pocket change and receipts
in rooms made for leaving.

Cigarettes get huffed
in the bathroom while
a fan whirrs so that
these places can go on

without any imprint
of what anxiety,
anguish or lust
takes place there.

Rivers of movement
pass through hotel rooms.
Most lodgers drive off
down highways that are

themselves rivers
cut through countrysides
teeming with distinct histories,

if anyone wanted to stop
and look. No one does.
But at least interstates
have destinations.

Hotel rooms stay behind,
ice slabs in a wide,
broken-up river
of indifference.

WOMAN HOLLERING CREEK

I pass over "Woman Hollering Creek" headed East on I-10
at 80 mph on a rainy Saturday and wake up Sunday
morning with an image of a woman with matted black hair
on the banks alone, wailing about what, I do not know,
but loud and long enough that half the town or more

must have stopped to listen. Maybe her grief was too sharp
for anyone to want to intrude, or maybe the better part
of that county was someways related to the woman's husband
and just couldn't afford the complex burden of opposing him
to give her sympathy, risk years of his crosswise looks, curses.

A lot of those Lone Star buckeroos are hard-cooked
sons of bitches. So she stayed down below the flat, level land,
its cattle, its sagebrush, mesquite, in a deep cut where rains
run off. Sand, caked mud. She screamed, wrenched.
The town listened, and remembered. Talked about it for years,

even after she left, went back to where it was she came from.
'Til finally she came to own that creek with her sorrow.
Still does own it by name in a land so dry, so hostile
that many of the people there had given up on having feelings
other than thirst or hunger a few generations beforehand.

THEY SAY THEY SEE THEM

Over and over they tell each other
        the how and when
of mighty angels,   that materialize
on night streets   out of neon.

Drawings and stories get swapped
between street kids: New Orleans, Seattle,
Dallas, Buffalo, Miami, Minneapolis.
Ones we see huddled along tracks,
down alleys, wearing slept-in shirts.

Teens whose practiced eyes
grade meth quicker than a cop's.
Some have had rich-guy-cock
shoved down their throats since
they were the age when my children
still gamboled over soccer grass.

Over and over they tell each other
        the when and how of
ENFORCERS,   PROTECTORS.
Spectral angels that blow up from
        the cold depths
        of subway tunnels.

Midday, burning time, hungry,
they spy great Archangels lounging
in clusters,   laughing, gabbing
on the ladder works behind billboards
while interstate motorists hurdle by.

Above Natchitoches on I-49

Cumuli curdle, skim pan-flat,
press full weight upon an atmospheric layer

and cast shade that brindles the asphalt
I rocket over at seventy-five miles per hour.

At shadow's edge I see how high wind lashes above,
sense quick exchanges, how gases churn.

In this way I pass into the pure, aloof world
of imagination for a half-mile or more.

Snap back to a straight-edged highway
that pulls me along its southbound track.

This behemoth roar of gear, gnash, rip
parts one hundred twenty miles of solemn forest.

Caddy Daddy

for Marty Marak

Marty up in Shreveport finally got himself
his own "cashmere brown" Cadillac
to promenade under live oaks. To park up front
in the church lot before he hobbles in.
Marty's got a real bad right knee.

He passes parishioners, leans a little forward,
steadies himself with a grip of the left hand
along the hemline of his sports coat,
spins, pivots into the front left row
which is one of two rows that are the only ones
in all of Saint Agnes Catholic Church
where there is no kneeler. Chooses that row
because he cannot genuflect.
Not on that bum knee.

So Marty tools his cashmere brown tank
up to the bank teller's drive-through,
over to his strip mall office, around to
the take out line at Popeye's finger-sucking-good
chicken shack. Marty has started to call himself –
has deemed himself – "Caddy Daddy".

This blithe spirit wrapped in courtly manners
has always been a business guy. But Marty used to
chug around Shreve's-town in a second hand Yugo.
No mistaking him back then. Yugos came
in military issue colors: olive, coal, khaki, pigeon-gray.
At best, piss-water yeller or a faded,
what's-left-of-lipstick-red-come-supper-time.

Marty flat doted on his Yugo.
How it never failed to start.
But that was back when he
really needed to get places,
had so much to do, instead of largely
to just let the whole city surmise
how clear it is that Marty has arrived.

CRACKING IT OPEN
a response to "Parking Lot Poem"
by J. R. Solonche

Read a guy who says a parking lot is damn near the best place
on this wide, wicked globe to pen a poem or tap
it into the most convenient device you brought because
sky yawns quasi-eternal above window glass and slab.

Sky fuller there than what glazes roof angles from a Paris garret
or what sloops over Irish Channel wave-gnash
viewed from the stone works of a winded tower.
Then he touches on how sun rips along razor wire on any
weekly fifty-minute hour spent in a prison yard.

Which this guy reckons to be the three best places anywhere
upon this Google Earth we crawl across like crickets
if you want to escape into blue or poetry.
A parking lot is his selection for fourth best
per its vast aurora set to plumb depths of boredom.

Immediately I leap up to add that I'd like to, just once,
write while looking onto Washington Square –
my window laced with plumes of smoke
as kids wait for their bus in coats way too big,
loose and open as they scoop, dash, duck
and toss gray-ashed snow. All the while thinking
I'd have to be a millionaire to afford more than
two swatches of sky gaze anywhere in Gotham town.

What about the excellence of me, settled down
in a suburban home apart from chain saw whines
on one more solo journey through yet another Thursday
beside a bulldog whose ribs give timbre to a snore
(deep, oceanic) that seems to take the whole sky in,
let it out. But slow, with utter pleasure.

CARS OF MEMORY
AND FORGETTING

Others Have the Appalachians, the Rockies; We Have a River

for Renee Stickels Ruzicka

You always said I treated my black sedan
like it was a truck. I did.
I had a rack for the roof
so the canoe could ride snug
as we raced up above New Roads.

A friend had an egress road
through a cow pasture on the levee.
I'd goose it to take a plunge down
over grasses still sopping wet with dew.
I'd dive down, run alongside long pools
that collect and stay after Spring floods,
hold turtles, bull frogs, moccasins, bream.

Then I'd stitch a pathway between
saplings, vines, ruts, broken limbs,
out to a stand of sycamore and willow
that line the banks on up to Minnesota.

One more perilous 45 degree dip to get
to a level ledge along water's edge
where I could park on a caked mix
of clay, gravel, sand; load the canoe.

We'd aim our bow into a deceptively gentle ebb,
paddle a half mile to an island with sand
spread wider than most city parks. Terns,
egrets, firewood, silence – except for wind
and the churn of an occasional tug.

I called that little beast of a car Zero
because I nabbed it for zero down,
zero interest, zero to pay for six months
but it was worth its weight in truck power,

seemed tiny left alone on a spit of shore
when we saw it across the Mississippi's flow
from a vast stretch of sand beach
that sang in our hearts, and still does.

WE BROKE DOWN
for Renee Stickels Ruzicka

The first time I know I have a good woman
is after a tire blows one Sunday morning in April.
Aqueous mirages waver over concrete
as eighteen-wheeler back drafts
rock my car on its shocks.

We are in the emergency lane
on a long stretch over marsh water.
Concrete slabs supported by pilings
quake as the big trucks blast past.

I have to get on my back, work
the spare from under the chassis.
I drop to my knee to set the jack
into a neatly designed buttress on the frame,
then I jerk my full body weight down
to break lugs welded to wheel metal.

Finally I notice you up stream in the emergency lane.
You lean out, glare at oncoming traffic.
In the face of that fierce demeanor
trucks and cars steadily skirt to the far lane.

I will never have a photo of how fervently
beautiful you were then, raked by wind,
backed by the wide blaze of Gulf blue.
Still, it will become the image by which I know you,
truer than any lens could ever capture.

ELK

Nothing but night and high desert asphalt.
An oncoming car flutters its headlights
just before I hit something big in the road,
absorb the jolt through tires and struts.
Impact, spasm. My Kia flies over a limp body.

I see the oncoming driver slow. In my rearview
the van circles, cuts back. Headlights show a form the size
of a yearling filly down in the road. I turn around, join him.
He says that the now inert elk just barely clipped
an aluminum ladder to his roof rack at the back
of his van. Aluminum got ripped to shreds, rattled off.

Pickups, eighteen-wheelers tear past.
Sharp valley wind, stone cold in August.
"If you've got rope we can get it out of the lane."
All I have is a coil of heavy-duty electric cord.
I tie the fluorescent orange cord around the bones
of the cow's back legs. First pull, no budge.
"Four hundred pounds probably," he says.

"Go on three," I say, count. The cow's still warm
hinds rise, slide a few inches. "Trevor. Trevor!"
he yells. "Get out here." With the three of us,
his kid maybe fifteen, we get it going seven inches
at a tug until that heavy cord breaks. I go
sprawling, my glasses fly onto emergency
lane asphalt. I bang one hip up good,
scrape an elbow. Then I retie the legs.
We go back at it, half a foot at a go
'til we get that carcass to the side,
sixteen hard inches off the rip of traffic.

In our Sorento, my wife is done crying. I
think of how those hinds sailed the elk
over brush in whisks and steady arcs less
than a half hour ago. The soft muzzle that
took in shrub grass, berries, lowered to nestle
upon the flanks of her young. But when
you're dead in the mountains, you're just
heavy dead, and I have a lot of asphalt ahead.

Phantom Canyon Road

Sickle sweep, dip and lift along State 50 west from Pueblo.
Right at a light at the corner of Diamond and Empty.
Unwind down four miles of asphalt. Past a landfill
fronted by sheets of tin, you'll round a wall of red rock
where Phantom Canyon lifts its shoulders. Runs thirty
tortured miles beside a gulch. The road was first upaxed,
tracked and trestled by Coolies and scruffy bastards
in the 1860s – product of the Pike's Peak gold rush.

Remote. Mule country. Even the squirrels look burly
with enough bristle to start a cap. Hares, sidewinders.
Elevator-size blocks thrust up in tumult. Crashed,
arrested, breathless, once-violent eruptions of rock.
Waves of rock. Strata. Brick-colored boulders
totter on cliff ledges or rest below in the arroyo
that curls at the base of it all – wrenching testament to
water's force. I can tell you this, the winds are fierce.
High mountain air turns unstable. Splits great forks
of lightning, wild, rampaging strikes. But usually all
is calm, naked, empty. As if nothing has ever happened
in a place too forsaken for hope. Sage, dun, salmon, dust.

Countless millennia of dust shifts under the scythe
of buzzard's wings. Isolation in the dead stroke of day
poised to give over to a riot of creep and trot in a dark
that brings this canyon to life. Night of wolf, bear, bobcat,
elk, moon. If you go, I want you to bring back something I left.
For me. You will find it wedged between two boulders just past
the single-lane tunnel in which I could taste various minerals within
that chiseled rock, the way a rattler darts its tongue to know
what the air carries. What I want back, if you go, what I need again
from this place where what has not been crushed rises, is please –
bring awe back to me. It is there, everywhere. Awe.

Postcard from the Amari Valley

You can't breathe and get by any oncoming vehicle
and hold the road all at the same instant for the treachery
of goat paths turned into roads near Frati.

Beside bramble, wattle and daub,
lamb and cattle huddle.
Vistas galore as lightning
licks chasms and crests
that were the first go of Crete.

Ancient Spilli holds one church
whose icons blaze gold,
and a score of shops
within its zigzag maze.
In Silli's marbled plaza
a spring spouts from the mouths
of twenty-five lions sculpted back
when the Minoan of Knossos
ruled this end of the Mediterranean.
Locals come up, bend forward,
fill plastic jugs.

We drive on into even more jagged terrain.
Stop at a well lit bar in a village
suspended on a cliff's edge.
Wizened and draped in black,
a grandmother fixes sharp eyes on us.
Oak tables, oak chairs, men in blue jeans
or suit coats smoke and chat.
Lightning blows the sky apart.

To Savor

Driving back from Colorado in the Spring
my youngest daughter had a tin of slippery elm lozenges.
Somewhere in Oklahoma where they had more signs
to cemeteries than signs to towns

we decided to see who could keep the disc
from dissolving for the longest.
Just let that host rest on our tongues
as fence posts and tall weeds whizzed by.

It is June now. I am back in Colorado.
This time with desks, mattresses and couches
stacked in a U-haul to bring my other daughter,
the eldest, back home where she can start again
with her old name. Leave that other name behind
in the scorching forest fires of a Colorado summer.

The first time my eldest daughter moved it was to Portland.
Catching views of Mount St. Helens over and over
we drove an insane labyrinth of one way streets
around an Ikea store for over an hour.

We could not get out of the maze to the store.
We raged, swore, doubled-over with laughter.
All to get furniture we are taking home again today.

At times you build your life with instructions typed by someone
who doesn't understand your language well. You keep turning
the diagrams different ways in different lights to figure it all out.

That is the way it is between memory and moment.
One second I try to solve myself, solve my daughters.
The next I just want to savor us together again.

Giving the Mitsubitchy
for Meagan Guidry

Meagan "sold" that "Mitsubitchy"
      bald-black, treadbare tires
for nothing but an aria
to help out a busboy who had been
      riding bus lines
      an hour each way to have
to the honor of picking up plates
ate off by the high commissioner
      of "What Where",
ate off by women
      think they so foul smell
they gotta douse themselves,
      douse their lady parts,
with eau de dis, eau de dat

'til they are not they,
not to Meagan as she waits their
      tedium and whim
not to this busboy who can't
wait until his fresh young wife
picks him up in the rattle trap
      he got on a promise.

Meagan even allowed him to pay on time,
         pay as he can,
on accounta the big heart Meagan has,
wanting to help always
whoever. This what makes her
so valuable to the whole enterprise
that they can afford her
a whole lotta minimum wage.

One guy in the kitchen sees something
on his off day and everybody finds out
that the reason the wifey
be driving busboy daily in that "Mitsubitchy"
         to the place he sweat and scamper
is that she be driving herself daily
         to a rendezvous
         with her ex-boyfriend for a little yum-yum.
So all those plates come tumble down.
That Mitsubishi rumbles off to Arkansas
leaving Meagan with nothing but more reason
to jot down order, after order, after order.

Dawn's Breath

I put my nose into a morning soft as biscuit dough
under the floured-up hands of a baker.

In nearby houses, wives snug bosoms into bras.
I shove my foot through the car door frame.

Going to get the engine hot. Going to goose the thing
despite a metallic taste on my tongue thick as memory.

Pickups drag past mailboxes, tossed cups, dew damp weeds
and a tree line sketched at the back of a pasture.

Strands of wind draw clouds out thin like they were formed
by the violin strokes of a child that does not really want to practice.

Though I cart old problems, stubborn issues
up the same old road as yesterday, my hood sails

into the abundant, over spilling fog of seven a.m.
At a light I idle, swathed in clouds of sound,

then gas off into this city's burgeoning surge.
Dawn's breath whistles in the window.

On a Tuesday
Before Trump Meets
with Kim Jong-Un

Under towering cumulonimbi,
a garbage truck grumbles.
Fluorescent Nikes flick by.
Neighbors pass, noses in their phones.
Prismatic sheens of motor oil
stain asphalt as dawn casts
spare change up and down the block.

The Daily lets me know that diabetes
enjoys another decade of record growth.
Millions of children have greater pangs of hunger
than school lunches can begin to address. Stores
are loaded with everything imaginable.

Frustration is palpable on talk shows, in editorials
and during the glacial movement of traffic.
The gates of gated communities glide effortlessly.

Under a nearby bridge a man we know as "Teddy"
wakes to grasp the handle of his shopping cart
and lumber off toward the Quick Mart
where soccer-dads pump petrol.

Salesmen in their bathrooms
toss back Listerine, gargle with full bravado,
as if this city holds the talented throats
of a thousand celebrated tenors, each of them ready
to stroll out onto life's opera stage.
They brace themselves by their back doors,
dash through a sudden, blistering rain
with magnificent abandon.

Missiles point to the East.
Missiles point to the West. Whales drift
past atomic submarines. I drive to work
taking the usual route. Dawn bursts
into its ordinary, raucous crescendo.

Often When

half-light furred on concrete,
I get out in the hospital parking lot,

I find someone still sitting
in a car parked near mine

as they text, comb hair,
listen to a radio announcer

and taste a quiet instant
before they go in to care for others.

As I Leave the Hospital

The air in the parking garage
is soft and moist like breath
from the muzzle of the old mare
that used to follow at my elbow
as I carried her bucket of oats
across the paddock.

Many mornings as I did chores,
our mare emerged from fog
to trail me, long neck bent.
Its gentle nuzzles come back now
as I turn my Kia's key, rev up.
Glide slowly down concrete ramps
shadowed gray at dawn, gray
at midday, gray at dusk.

I ease onto Florida Blvd.
beside a Nissan pickup. A student
with a backpack on a bike
props on one foot at the red light.
I think for a second how
Mrs. Ethridge in Room 503
can barely reach the glass
I left on her over-bed table
less than a foot from her hand.

I am breezing along now.
All of the motor's horsepower courses
me down the corridor that leads
to my house where I will go
to my patio, sit as calm as a mare
in a white emulsion of fog.

OF A SUDDEN, STARTLED RIGHT

Four-door pickups mate in parking lots.
They must. Two thirds of our Tom Thumbs,
Dickless Harrys in this fire engine red state
tear through "pink" lights, halt traffic
to execute perilous U-turns, jump lanes
at jet pilot speeds in monster trucks
that shimmer chrome, glint primary colors.

You have to wait in queue while these
sons-a-bs shoehorn their outsized egos
into hospital parking slots.
Country music blares across the lot.

Damned if you don't run into them in a room
where they attend a fractured mother or crumpled wife,
tuck sheet and blanket over them with care.
They joke and yodel with the P. C. A.
as if humor could dispel terror.
They hang on the doc's words lost as kittens,
blink up from an impossible couch
where they just spent six cramped hours,
bones poised in an oddity of tender reverence.

Quietly Forward
for Renee Stickels Ruzicka

I've been with women who dabbed on makeup,
smeared lip gloss, pulled at split ends as I drove.
Ones that I talked to just to make an impression.
Women who thought the whole world
might want to hear what they say to their cell phone.
Ones that wanted a different station,
maybe a different driver.

Renee and I ride quietly.
We fly along as if we were
sitting side to side on the patio
not really moving at all.

Trees whose roots mix together
below the forest floor exchange birds,
squirrels, nutrients. Accept the wind,
rain, seasons. Multiply leaves.
Translate sunlight into oxygen.

DELAYED GRATIFIXATION

Tricked out T-bird,
convertible roof folds
behind that nook of a seat,
where you stick the tag-along-friend,
all elbows, knees and envy.

Four on the floor. Mooneyes white wall tires.
Lipstick red seats. Headlamps make
sweet eyes beside lightning bolts of chrome.

Pat's been retooling Classic cars since
his youngest son died. He just walked away
from his law practice. Says that it eases
his heart to rub, sand, paint fenders.

Told me that his uncle owned a Sinclair station
where he started pumping when he was fourteen.
There was a Catholic girl who glided in on Wednesdays
after tennis practice, racket slung in the back.
Came in the same model Thunder-bird
that Pat is showing to me now.

Every week Pat cleaned her window glass
as she spread her legs apart to let him
glimpse white inside her thighs.
Pat says his elbow went half crazy
trying to make her world shine
the way she wanted it.

That's the way his T-bird looks –
lean, star-sexed.

Sunday

I

The car of Sunday
is content to stay where it sits.
To set its shadow onto cement,
let breezes fuss over hood and roof
like hands making up a bed.

Sunday cars are the cars of contentment.
Their trunks reflect towers of cumuli,
breathless blues where jets prowl.

II

When a Jaguar or a midnight blue sedan sit idle
there is still an animating agent
in how wind bellies down on the hood,
slides along svelte swerves, swirls
in ribbons off the back.

The two of them – contour and wind – are like old lovers
who come together again and again
over centuries, over eons.
Early on they discovered each other
as wind raced along the flanks of a horse
escaping a canyon that was on fire.

Over and over old lovers come together.
Wind and sand dunes. Wind rippling the river.
Wind and contour change places, with wind
sometimes born as the woman, sometimes born as the man.
Contour, surface sometimes the man, sometimes the woman.
That is the way it is for old lovers, always the same, always changing.
Charged and alive even at rest – fenders and wind
and old lovers on Sundays.

III

The universe spins in a sort of expanding, contracting hurricane,
a nautilus, a double helix. Things on the left
get carried to the right, then left again,
each held roughly at the same distance and arrangement
toward the other, though tumbled into various postures
as they slip down the sleeve of a vortex,

only to be shot back out again. Out of the eye
at the bottom of all this violent twisting and becoming.
Death and rebirth happen when we are not fully here.
Death and rebirth happen within everyone else's full view.

I got the order from the dispatch clerk at the parlor.
Isn't it odd how that formal word, "parlor",
that Victorian term, has always been reserved
for the awful holes-in-the-wall pizzas come from.
I got the order among others on what was the last
run of a night when this city beside a river
was in the grip of an immense rain
that swept up under the chassis of my coupe,
gushed in gutters, gurgled down drains.

It was my last stop. She answered the door.
There was a lamp behind her. Light bent
around her silhouette as she counted out the dollars.
We started to talk. It was happening again.
"This time she is the elder." I didn't think that yet.
That was months away and not really ever even a thought.
Yet it became something that pierced me months later
after we made love again completely. Fell back. Then
came together. "This time she is the elder," my dream said
though I could not hear it, wrapped in her skin as I was,
the hurricane of her hair spread across the pillow.
Two weeks later I shipped out to the Pacific.
Guam became my last stop.

Born over and over, contour and wind, man and woman
the same yet different. Fresh, never quite sure
what has happened before, what is to come.

V

There was the time we met at the cistern.
The time I was in a coach and four and spied you
coming across a field wielding a scythe.
That was the time we would have ended up in America
if you hadn't died of scarlet fever in the bowels of a ship.
There has always been a wildness of the sea
that flows back and forth between us.

VI

It is Sunday. I am in a hammock
wrapped in the sort of God-awful heat
that seizes old arteries, pounds their fibers till blood
can burst its walls, flood down pathways
never meant for the coursing of blood.

The apostle Paul was struck down by light.
Many tribal villages revered their epileptics, their afflicted
because they realized that visions often rip into view
in the middle of seizures and strokes.

Last week a patient told me that he fell down,
was rendered speechless. That he remembered
being rolled on a gurney, was vaguely aware
how doctors and nurses attended to him
though he was barely there.

He was wheeling through the wheeling stars.
My patient was in that sort of struggle he never
wanted to be in but knew he would be someday.
His quiet was not quiet but he
came back filled with a new quiet.

VII

My quiet in a hammock on a Sunday
is quiet. My stillness is stillness. So much
stillness and quiet that I am immense.
Immense inside a heat so great it could burst.
All of these leaves above me are emblems.
Winded branches are the rattling tongues of our joy.

Sometimes we talk in the kitchen or on the patio,
try to decide, decipher, remember, fix
exactly how many years we have been together.
Sometimes we sense that it is not years but lifetimes
upon lifetimes that we have come together under leaves
and branches that are the spun veins, the fluted tendrils of joy.

VIII

Even on a Sunday someone may wander out,
turn the key, head off toward a store.
But still, Sunday cars roll out
like they have forgotten
what to do with themselves,
like infants twisting and turning without purpose
except to twist and turn without purpose
in kingdoms of unmeasured time
and wondrous forgetting.

Acknowledgments

The following poems have appeared previously in print and / or online:

Bad Work was published in Capsule Stories

Chevy Pick-up, Loaded was included in the anthology By the Light of a Neon Moon, published by Madville Press

Elk can be seen in the pages of El Portal Journal

Go, Just Go first appeared in Capsule Stories

Hell of a Ride was among a number of mine released in The Poet

I Hitchhiked All Over was first published by Capsule Stories

I've Had a Lot of Beaters was placed online in Rat's Ass Review

Others Have the Appalachians, the Rockies; We Have a River was included in Bosque Review

Phantom Canyon Road appeared in Manzano Mountain Review

Postcard from the Amari Valley was published in The Poet

The Road to Alligator Bayou came out in Blue Mountain Review

Teen Ode first appeared in South Pedro River Review

When Cars Were America appeared in Pride 7 Deadly Sins Vol. 7, published by Pure Slush Books

When I First Heard 'Black Magic Woman' came out in The Stillwater Review

Woman Hollering Creek (published as I Pass Over) was published in Bosque Review

About Ed Ruzicka

Raised beside creeks and cornfields not far from Chicago, at seventeen Ed Ruzicka followed his thumb to the Sunset Strip in Los Angeles and later to Harvard Square, Bread Loaf at Middlebury, Vermont and Juarez, Mexico.

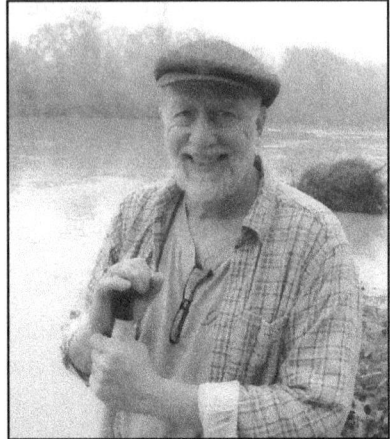

In 1973 Ed hitchhiked to Louisiana to visit Charles N. deGravelles. He saw live oaks and never left. For decades, Ed has practiced and continues to love practicing occupational therapy. He first worked with special needs children in a local school system but now works part-time, primarily in rehab settings.

Ed has previously published one full-length book of poems, *Engines of Belief*.

Ed's poems have appeared in the *Atlanta Review*, *Rattle*, *Canary* and the *San Pedro River Review* as well as many other literary journals and anthologies. Ed has been a finalist for the Dana Award and the New Millennium Award.

Ed lives with his wife Renee and Tucker the doddering bulldog in Baton Rouge, Louisiana.

Find more about Ed and his work at underlined edrpoet.com.

ABOUT CHARLES N. deGRAVELLES

Charles N. deGravelles is a writer, composer and graphic artist.

His poems and stories have appeared widely in literary magazines and anthologies. His biography of football icon, felon and prison reformer, Billy Cannon, won the Louisiana Library Association Best Book of 2015.

He is a deacon in the Episcopal Church and a social justice advocate who has worked with and for the incarcerated, the addicted, the displaced and the homeless.

Photo by Tim Easton

ABOUT TRUTH SERUM PRESS

Established in 2014, Truth Serum Press is based in Adelaide, Australia, but publishes books from authors in all parts of the English-speaking world.

Truth Serum Press (along with sister presses Pure Slush Books and Everytime Press) is part of the Bequem Publishing collective.

Truth Serum Press has published novels, novellas, and short story collections; as well as poetry anthologies and collections.

Similarly, when the mood strikes us, we publish multi-author anthologies. Generally, we publish fiction ... and sometimes (just sometimes) we publish non-fiction.

We publish in English, and we would gladly publish in other languages if we understood them.

We like books that take us to new places, inside new minds and hearts, and to new experiences.

We also like to laugh.

Visit our website at https://truthserumpress.net/.

Also from TRUTH SERUM PRESS

truthserumpress.net/catalogue/

- *Indigomania* Truth Serum Vol. 4
  978-1-925536-03-4 (paperback) 978-1-925536-84-30 (eBook)
- *Stories My Gay Uncle Told Me* Truth Serum Vol. 3
  978-1-925536-86-7 (paperback) 978-1-925536-87-4 (eBook)
- *Wiser* Truth Serum Vol. 2
  978-1-925101-31-7 (paperback) 978-1-925101-32-4 (eBook)

- *True* Truth Serum Vol. 1
  978-1-925101-29-4 (paperback) 978-1-925101-30-0 (eBook)
- *A Short Walk to the Sea* by Eddy Knight
  978-1-925536-01-1 (paperback) 978-1-925536-02-7 (eBook)
- *How to Catch Flathead* by Peter Michal
  978-1-925536-94-2 (paperback) 978-1-925536-95-9 (eBook)

Also from TRUTH SERUM PRESS

truthserumpress.net/catalogue/

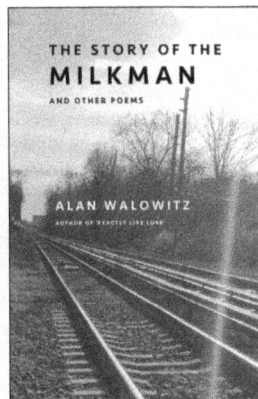

- *The Last Free Man* by Lewis Woolston
  978-1-925536-88-1 (paperback) 978-1-925536-89-8 (eBook)
- *Filthy Sucre* by Nod Ghosh
  978-1-925536-92-8 (paperback) 978-1-925536-93-5 (eBook)
- *The Story of the Milkman* by Alan Walowitz
  978-1-925536-76-8 (paperback) 978-1-925536-77-5 (eBook)

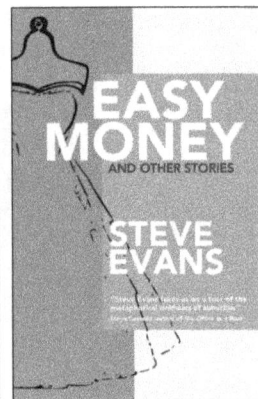

- *Minotaur and Other Stories* by Salvatore Difalco
  978-1-925536-79-9 (paperback) 978-1-925536-80-5 (eBook)
- *The Book of Acrostics* by John Lambremont, Sr.
  978-1-925536-52-2 (paperback) 978-1-925536-53-9 (eBook)
- *Easy Money* by Steve Evans
  978-1-925536-81-2 (paperback) 978-1-925536-82-9 (eBook)

Also from TRUTH SERUM PRESS

truthserumpress.net/catalogue/

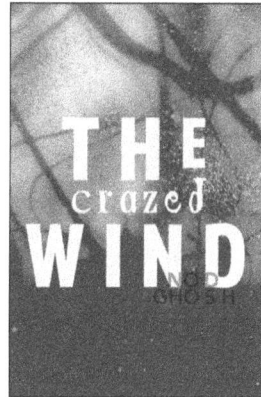

- *Square Pegs* by Rob Walker
  978-1-925536-62-1 (paperback) 978-1-925536-63-8 (eBook)
- *Cheat Sheets* by Edward O'Dwyer
  978-1-925536-60-7 (paperback) 978-1-925536-61-4 (eBook)
- *The Crazed Wind* by Nod Ghosh
  978-1-925536-58-4 (paperback) 978-1-925536-59-1 (eBook)

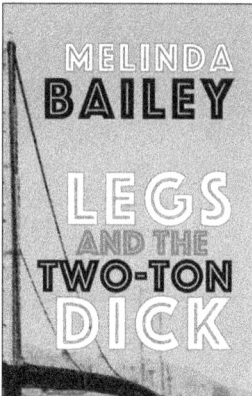

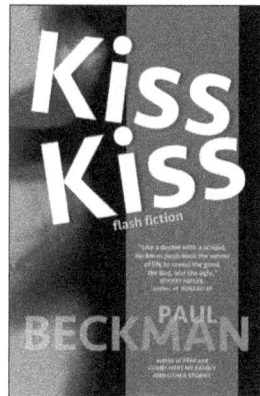

- *Legs and the Two-Ton Dick* by Melinda Bailey
  978-1-925536-37-9 (paperback) 978-1-925536-38-6 (eBook)
- *Dollhouse Masquerade* by Samuel E. Cole
  978-1-925536-43-0 (paperback) 978-1-925536-44-7 (eBook)
- *Kiss Kiss* by Paul Beckman
  978-1-925536-21-8 (paperback) 978-1-925536-22-5 (eBook)

Also from TRUTH SERUM PRESS

truthserumpress.net/catalogue/

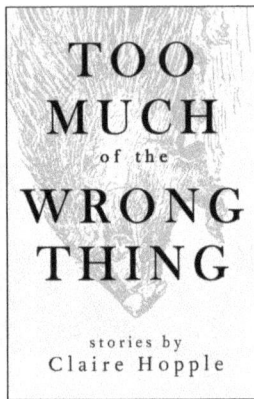

- *Inklings* by Irene Buckler
  978-1-925536-41-6 (paperback) 978-1-925536-42-3 (eBook)
- *On the Bitch* by Matt Potter
  978-1-925536-45-4 (paperback) 978-1-925536-46-1 (eBook)
- *Too Much of the Wrong Thing* by Claire Hopple
  978-1-925536-33-1 (paperback) 978-1-925536-34-8 (eBook)

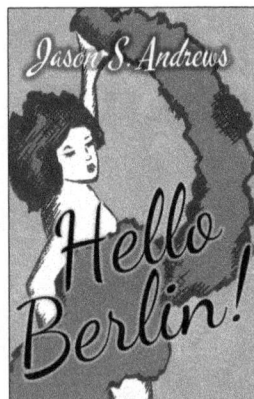

- *Track Tales* by Mercedes Webb-Pullman
  978-1-925536-35-5 (paperback) 978-1-925536-36-2 (eBook)
- *Luck and Other Truths* by Richard Mark Glover
  978-1-925101-77-5 (paperback) 978-1-925536-04-1 (eBook)
- *Hello Berlin!* by Jason S. Andrews
  978-1-925536-11-9 (paperback) 978-1-925536-12-6 (eBook)

Also from TRUTH SERUM PRESS

truthserumpress.net/catalogue/

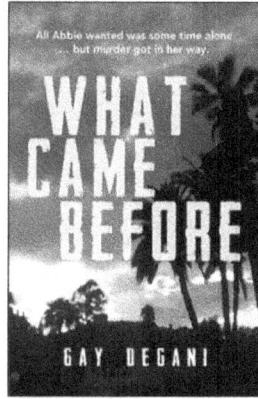

- *Deer Michigan* by Jack C. Buck
  978-1-925536-25-6 (paperback) 978-1-925536-26-3 (eBook)
- *What Came Before* by Gay Degani
  978-1-925536-05-8 (paperback) 978-1-925536-06-5 (eBook)
- *Rain Check* by Levi Andrew Noe
  978-1-925536-09-6 (paperback) 978-1-925536-10-2 (eBook)

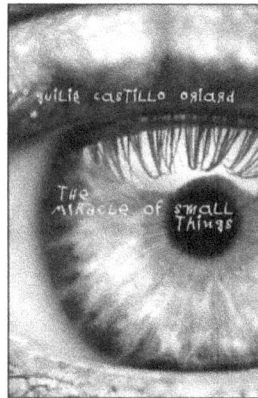

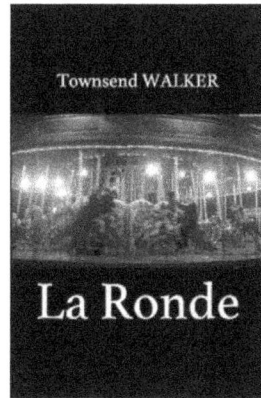

- *Based on True Stories* by Matt Potter
  978-1-925101-75-1 (paperback) 978-1-925101-76-8 (eBook)
- *The Miracle of Small Things* by Guilie Castillo Oriard
  978-1-925101-73-7 (paperback) 978-1-925101-74-4 (eBook)
- *La Ronde* by Townsend Walker
  978-1-925101-64-5 (paperback) 978-1-925101-65-2 (eBook)